I0814258

LET'S CODE!

THE BASICS OF CODING

BY SAMANTHA S. BELL

CONTENT CONSULTANT
DAVID A. LASH
ASSOCIATE PROFESSOR OF COMPUTER SCIENCE
AURORA UNIVERSITY

Kids Core
An Imprint of Abdo Publishing
abdobooks.com

abdobooks.com

Printed in the United States of America, North Mankato, Minnesota.
102023
012024

Cover Photo: Shutterstock Images
Interior Photos: Shutterstock Images, 4–5, 7, 8, 17, 18, 20–21, 23 (background, knight), 23 (dinosaur, fire), 28 (top), 28 (bottom), 29 (bottom); Andrey Popov/Shutterstock Images, 10; Ground Picture/Shutterstock Images, 12–13; Kayla Oaddams/WireImage/Getty Images, 15; Goroden Koff/Shutterstock Images, 24, 29 (top); Pisa Photography/Shutterstock Images, 26

Editor: Katharine Hale
Series Designer: Katharine Hale

Library of Congress Control Number: 2023939663

Publisher's Cataloging-in-Publication Data

Names: Bell, Samantha S., author.
Title: The basics of coding / by Samantha S. Bell
Description: Minneapolis, Minnesota: Abdo Publishing, 2024 | Series: Let's code! | Includes online resources and index.
Identifiers: ISBN 9781098292737 (lib. bdg.) | ISBN 9798384910671 (ebook)
Subjects: LCSH: Coding theory--Juvenile literature. | Computer programming--Juvenile literature. | Computers and children--Juvenile literature. | Computer programming--Juvenile literature.
Classification: DDC 005.1--dc23

CONTENTS

Technology can be a helpful tool when cooking.

CHAPTER 1

CODING WITH DINNER

Lydia's dad was driving her home from school. He pressed the button on the garage door opener. As the door opened, he pulled into the garage. When they got out of the car, he hit the lock button on his key fob. The car beeped.

"We should get started on dinner," Lydia's dad said. "Do you want to pick a recipe?"

"Yes!" Lydia had borrowed an ebook about cooking from the library's website. She was excited to find a recipe to try. Lydia ran into the house. She wanted to set the mood for cooking. She called out to her family's **smart** speaker. "Play our cooking playlist, please!"

"Playing Lydia's cooking playlist," the speaker replied.

Lydia started dancing to the music as she picked up her tablet. She swiped her finger across the screen to turn the book's pages. "Dad, how about lasagna?"

"Lasagna sounds great! What do we need?" her dad asked.

Smart speakers can help people cook by setting timers, answering questions, and more.

“Step one is preheating the oven,” Lydia said. Her dad pushed some buttons on the oven. It began heating up. Then they started on the rest of dinner.

Programmers write the code everyday devices need to function.

Reading an ebook on her tablet was one way Lydia used technology in her evening routine. The garage door opener, key fob, smart speaker, and oven all used technology too. They all required **software** to work. The software was created by coding.

What Is Coding?

Coding is the process of writing programs. Programs give instructions to a computer or other device. The instructions allow the machines to perform certain jobs. For example, a program made the garage door open. Programs caused the smart speaker to play music and the tablet to display the recipe. Coding created the programs that make these devices work.

Ways to Learn to Code

People learn to code in different ways. Many people study coding in college. Some people teach themselves. They may use **self-paced** classes online. Some people attend special coding classes called boot camps. There are even toys that teach people how to code. Robots are one example. People can learn code and use it to control a robot. The robots can create art, navigate mazes, and more.

There are many types of programming jobs. Programmers write in different programming languages.

Any device that involves a computer needs coding. The code is written by a programmer. The programmer writes the instructions for the device to follow. With the help of programmers and coding, Lydia and her dad made dinner by using helpful technology.

Explore Online

Visit the website below. Does it give any new information about coding that wasn't in Chapter One?

What Is a Computer Program?

abdocorelibrary.com/basics-of-coding

Programs on tablets, computers, and phones are written in code.

CHAPTER 2

THE JOB OF A PROGRAMMER

When a programmer writes code, they write the instructions for a device. The instructions are stored on memory chips or **hard drives**. Some instructions are written so a device responds to a person's command. Saving a document is one example.

When the user clicks a button, the computer saves the file. Other instructions let devices do tasks on their own. A **thermostat** is one of these devices. A person sets the desired temperature for a room. But the thermostat has to know when to turn the furnace on or off. It uses sensors and its coding to do this.

Some programmers work with large systems, such as factory machines. Programmers can also work with very small systems, such as wristwatches. Movies that use visual effects require coding. So do smartphones and tablets. Video games and websites use coding too. Coding creates the software that allows smart household items such as televisions and refrigerators to connect to the internet.

Shigeru Miyamoto created the *Super Mario* video game series. A movie based on the games was released in 2023. Lots of coding went into both the movie and the games.

Types of Programmers

Programmers usually work in teams. They often specialize in different areas. The most common programmers are web developers. They design the websites people use. Other programmers create apps for mobile devices, such as smartphones and tablets. They also keep the apps updated.

All Sides of a Website

There are three types of web developers. Front-end developers make websites look nice. They add graphics and animations. They also try to make the sites easy to use. Back-end developers build and maintain the code that runs the website. Full-stack programmers work on both the front and back ends.

Web developers create websites.

Video games also require teams of programmers. Some game developers create the systems that make everything work. Others work on the story and the art. Some build the **artificial intelligence** (AI) of characters a player meets. The programmers try to make the games fun and entertaining.

Many other devices run on software. Some programmers are called embedded systems developers. They write code for these devices.

Some cars use technology to sense vehicles around them. These cars can slow down or stop on their own to prevent crashes.

Cars are one example. Programmers create the software systems used in car technology, such as automatic braking. Programs are also used in traffic lights, smart TVs, and medical devices.

There are many other types of programmers. They all play an important part in keeping things running smoothly. Their work affects people's lives every day.

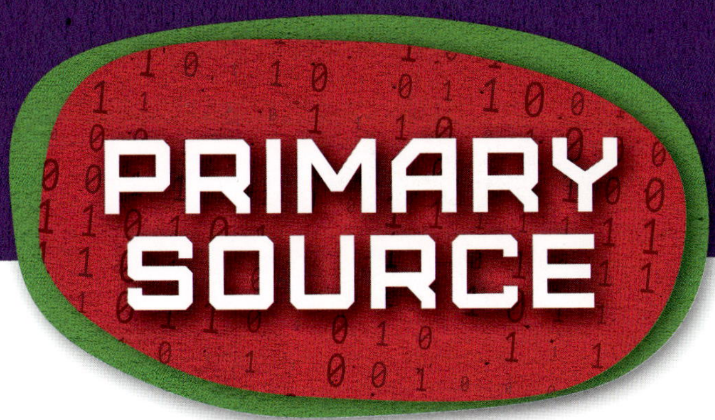

Hansel Lynn is the founder of the Coder School in San Diego, California. The school is open all year for kids who want to learn to code. Lynn said:

> For kids, learning to code isn't about the actual technology or coding languages. . . . Coding is about logic. To speak to a computer, you must give precise step-by-step instructions.

Source: Megan Woolsey. "Coding for Kids in San Diego." *San Diego Family,* n.d., sandiegofamily.com. Accessed 5 Mar. 2023.

What's the Big Idea?

What is this quote's main idea? Explain how the main idea is supported by details.

Programmers have to learn programming languages to communicate with computers.

CHAPTER 3

COMMUNICATING IN CODE

For a computer program to work, it must be presented in a language the computer understands. This is called machine language. Machine language contains only ones and zeroes.

Creating a program in machine language is not easy. The language is very challenging for people to write and read. Also, computers have different types of **processors**. The different processors require different instructions written in machine language.

To solve these problems, higher-level programming languages were created. These languages make it easier to write programs. They use words and symbols rather than just ones and zeroes. First, the programmer writes the code in a higher-level language. Then a program called a compiler translates the code into machine language. Now the computer can read and carry out the instructions.

How Code Works

```
if (knight.health > 0)

  if (getButtonPressed==Key.spacebar)
    Jump

  else if (getButtonPressed==Key.F)
    ShootFire
```

If the knight's health is above 0, the player can control it.

If the player presses the space bar, the knight jumps.

If the player presses the F key, the knight shoots fire.

This simplified sample code shows how code gives instructions to a program such as a video game.

Drag-and-drop programming languages help beginners learn how to code.

Lots of Languages

There are more than 9,000 higher-level languages. Some languages are used for only one type of job. Other languages can be used in several ways. Many computer programmers know more than one programming language.

One of the most common languages is called Java. It is often used for creating apps. JavaScript is used for making websites. Python is another language. It is often used by scientists and mathematicians. Game developers may use C++ or C#. Other popular languages include Swift, Ruby, and Go.

Drag-and-Drop

Some programming languages are called drag-and-drop languages. They use colorful blocks of code. The code is already written. Beginners can drag and drop blocks of code to create games, apps, and other projects. Some of these languages include Scratch, Blockly, and Beetle Blocks.

The everyday electronics people rely on use software coded by programmers.

Coding affects people's lives in many ways. From smartphones and laptops to traffic lights and security systems, coding helps keep things running smoothly. Coding creates the programs and software that make these devices work. Through coding, people can find even more ways to use technology.

Further Evidence

Look at the website below. Does it give any new evidence to support Chapter Three?

What Is Python?

abdocorelibrary.com/basics-of-coding

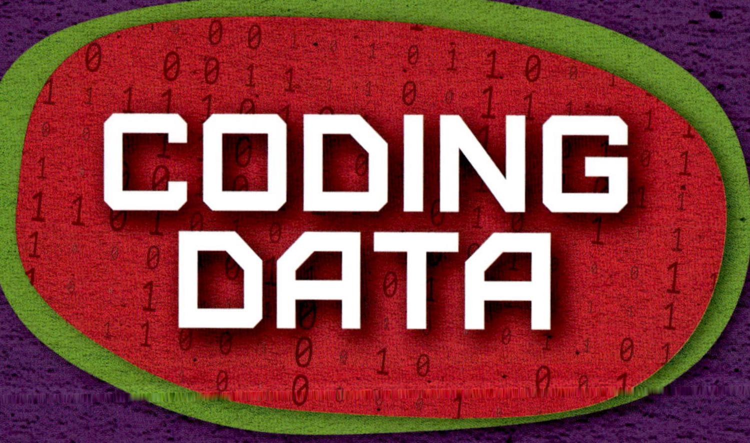

CODING DATA

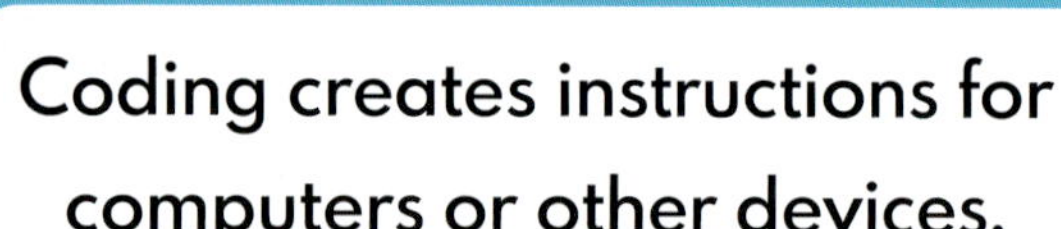

Coding creates instructions for computers or other devices.

Devices such as smart speakers run on programs coded by programmers.

Programmers write code in higher-level languages. The code is changed into machine language so a computer can understand it.

There are many different types of programmers. They work on websites, apps, games, machines, and more.

Glossary

artificial intelligence
the ability for a computer or machine to use information to learn and make decisions

hard drives
devices used by computers to store information

processors
the parts of a computer that carry out instructions

self-paced
designed to be completed at a student's own speed

smart
when referring to devices, describes appliances and other household items that can connect to the internet and other smart devices

software
all of the programs that give a computer instructions

thermostat
a device that controls the heat or cool air put out by a heating or cooling system to control the temperature

Online Resources

To learn more about coding, visit our free resource websites below.

Visit **abdocorelibrary.com** or scan this QR code for free Common Core resources for teachers and students, including vetted activities, multimedia, and booklinks, for deeper subject comprehension.

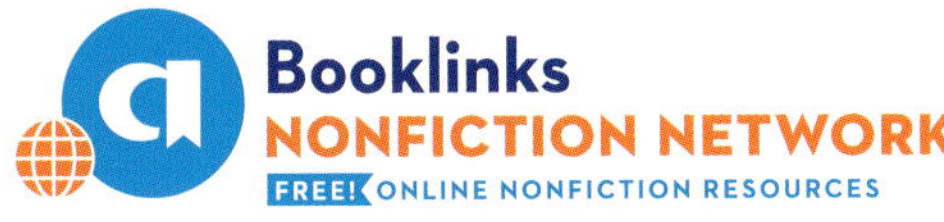

Visit **abdobooklinks.com** or scan this QR code for free additional online weblinks for further learning. These links are routinely monitored and updated to provide the most current information available.

Learn More

Bell, Samantha S. *Why Learn Coding?* Abdo, 2024.

González, Echo Elise. *Coding Languages.* World Book, 2021.

Woodcock, Jon. *Coding Projects in Scratch.* DK, 2019.

Index

About the Author

Samantha Bell lives in the foothills of the Blue Ridge Mountains with her family and lots of cats. She has written more than 150 nonfiction books for students. She learned HTML through building websites and wants to learn more programming languages.